Plastic
PINK FLAMINGOS

Rebecca Woodbury, Ph.D., M.Ed.

Gravitas Publications Inc.

Plastic
PINK FLAMINGOS

Illustrations: Janet Moneymaker

Plastic Pink Flamingos
ISBN 978-1-950415-15-1

Published by Gravitas Publications Inc.
Imprint: Real Science-4-Kids
www.gravitaspublications.com
www.realscience4kids.com

RS4K

Photo credits, Adobe Stock: Cover & Title Pg: SockaGPhoto; Above: Boots, Юлия Клюева; Flamingo, Loulou02; P.3. JenkoAtaman; P.5. Katie Chizhevskaya P.6. Irina Schmidt; P. 7. Loulou02; P.9. Boots, Юлия Клюева; Flamingo, Loulou02; P.15. T-shirt, airdone; Sweater, nys; P.17. ulga; P.18. teen00000; P.19. Pawel

Have you ever jumped in a puddle while wearing rubber boots?

Looks like fun!

Have you ever spotted
a plastic pink flamingo
strutting in the grass?

Silly
flamingo.

Do you know why rubber
rain boots are wiggly?

Do you know why plastic
pink flamingos are shiny?

Why?

Rain boots are made of **rubber.**

Pink flamingos are made of **plastic.**

Rubber and plastic are special molecules called **polymers**.

A **polymer** is a **molecule** with many repeating units.

"Poly" means "many."
"Mer" means "unit."

The word **polymer** means **many units.**

Polymers must be big molecules.

Sounds like it.

Review: ATOMS

Atoms are tiny building blocks that can link together.

Atoms make everything we see, touch, taste, and smell.

Review: MOLECULES

Molecules are made when **atoms link** together.

Polymers are like a long chain made of links.

- Each link in the chain is one unit.

- All the links together build the chain.

one link   **linked chain**

Polymer units are made of **hydrogen** and **carbon atoms**.

All the units together build a long polymer molecule.

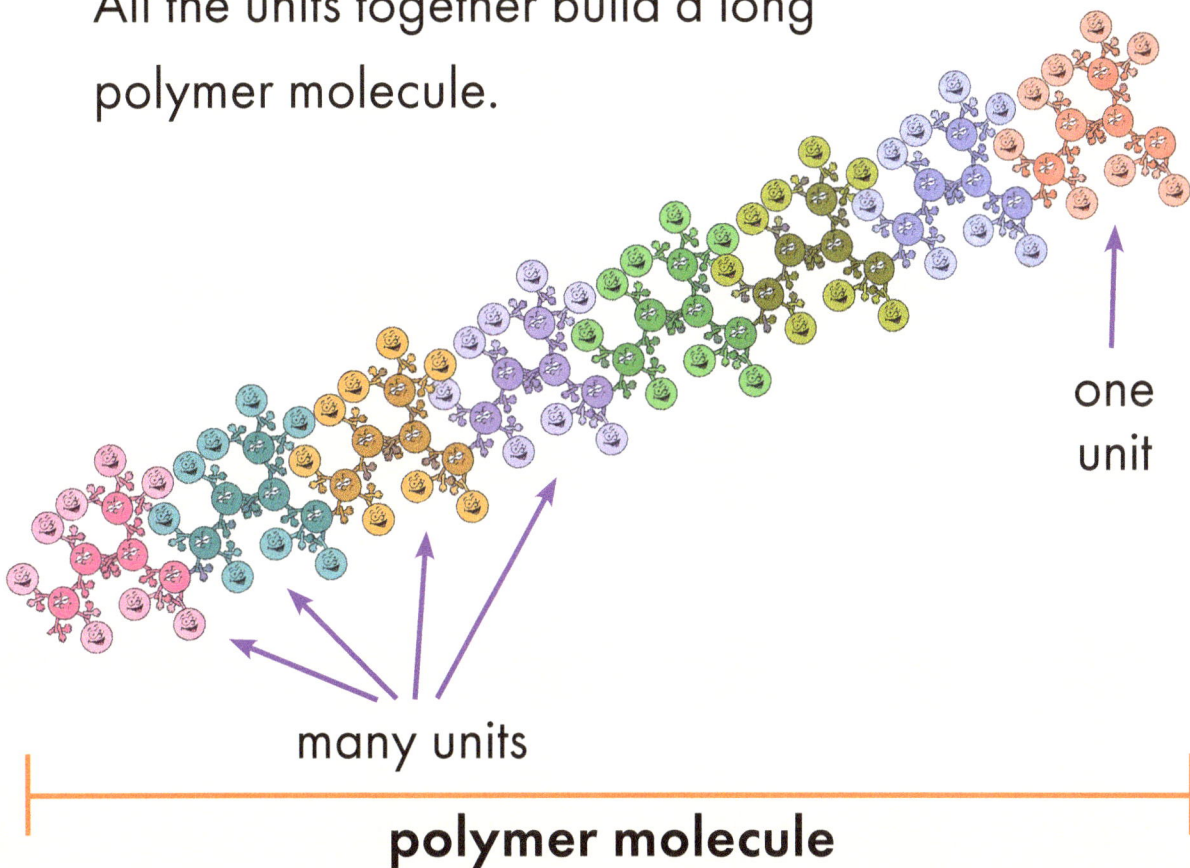

one unit

many units

polymer molecule

Polymers are used in many ways.

Styrofoam is a polymer used for packing.

Cotton fibers, **nylon**, and **polyester** are polymers used for clothing.

Is cheese a polymer?

I think so.

Plastics are used for bowls,
toy cars, and pink flamingos.

Polymers can change shape, color, and size.

Eggs have polymers that turn white when heat is added.

A rubber band can change shape and size.

Pull it back and watch it launch off your finger.

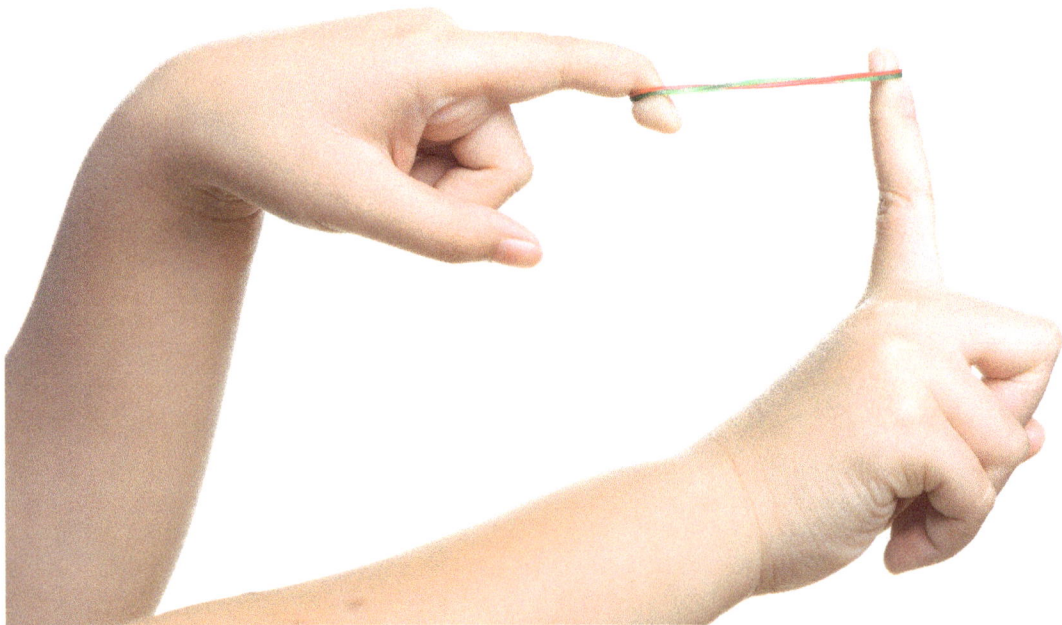

Glue is a polymer that can stop being sticky.

Add borax and it can be rolled into a ball.

Polymers are found everywhere.

rubber

plastic

polyester

nylon

plastic

cotton

rubber

Polymers are even used to make rocket ships!

How to say science words

atom   (AA-tuhm)

carbon   (KAHR-buhn)

electron (ih-LEHK-trahn)

hydrogen   (HIY-druh-juhn)

molecule   (MAH-lih-kyool)

nylon   (NIY-lahn)

polyester   (pah-lee-ES-tuhr)

polymer   (PAH-luh-muhr)

Styrofoam   (STIY-ruh-fohm)

www.ingramcontent.com/pod-product-compliance
Lightning Source LLC
Chambersburg PA
CBHW040151200326
41520CB00028B/7570